Carlos Calvet

The Background Field Theory

GRIN Verlag

Bibliografische Information der Deutschen Nationalbibliothek:

Die Deutsche Bibliothek verzeichnet diese Publikation in der Deutschen Nationalbibliografie; detaillierte bibliografische Daten sind im Internet über http://dnb.d-nb.de/ abrufbar.

Imprint:

Druck und Bindung: Books on Demand GmbH, Norderstedt Germany
ISBN: 978-3-638-65462-3

PART 1

EFFECTS AND EVIDENCE OF THE "BACKGROUND FIELD"

Carlos Calvet, "Effects and Evidence of the "Background Field"", Extraterrestrial Physics Review - EPR, Japan, Vol.1, No.3, March 2000

Carlos Calvet, "Effects and Evidence of the "Background Field"", Journal of New Energy, USA, Vol. 4, no. 4, Spring 2000

Author: Dr. Carlos Calvet
(Barcelona), Spain

Abstract

The Background Field (BF) is an advanced figurative quantum model of the Zero-Point Field, the probable origin of inertia, gravity and EM-fields. It explains also, why the speed of light is limited, and unclear phenomena like "antigravitation" and the "Tunnel Effect" by means of interactions between elementary particles and the BF. The BF fills up the whole universe and represents therefore a resistance to any moving particle, even to light.

Our universe consists of a BF located inside an absolute void space. We can imagine the BF as a 3-D matrix of virtual gravitons linked by strings. The tension of the strings produce the resistance, we know as "inertia". Contraction of the BF produces gravitation, while a spinning BF produces EM-fields.

If we made a "hole" in the BF, this hole would have no more any inherent resistance, thus allowing to increase the velocity of particles beyond "c". Such tiny holes in the BF are probably the origin of the "Tunnel Effect". This effect is the first evidence for the BF. A second evidence for the BF is the higher temperature of the solar corona (up to $2x10^{6}$°C) with respect to the photosphere (5,500°C) due to the high pressure of radiation that just leaves the material surface of the sun and produces large "holes" in the surrounding BF. The lack of resistance in such holes allows photons to accelerate beyond "c", thus gaining more energy and making increase the temperature in the corona.

Furthermore, if antigravitation resulted to be real, it would be the third evidence for this model, since it would be the result of *competition* between EM and gravitational fields for virtual particles derived from the BF. Experimental evidence of the BF would be the prediction that "c" increases in the outer space where the BF is less dense and inertia less intense. An anomalous behavior of gravitational attraction between celestial bodies could be the result of a hole in the BF between those bodies.

Finally, spacecrafts could use strong EM-radiation to produce holes in the BF and achieve practically unlimited speeds.

PACS-Number: 11.90.+t

Key Words: antigravitation, EM fields, gravitation, inertia, speed of light, strings, Tunnel Effect, virtual particles, Zero Point Energy/Field (ZPE/ZPF)

Introduction

There are several aspects in physics that have not yet been already understood, like the "Tunnel Effect" or inertia, may be because physics is lacking "something" or because the concept of "time" has not yet been completely revealed.

In modern physics, string theories are already describing particles as strings, and bosons as waves propagated by these strings. Parallel new developments in the understanding of the ZPE - that is the energy contained in the perfect vacuum - have been made recently, suggesting that in some way, inertia and gravitation are interrelated.

Inertia is probably the most unknown of all physical aspects. After failing to find an experimental proof of Mach's principle, that predicted that the gravitational effect of the whole matter contained in the universe is responsible for inertia, (Haisch et al.) suggested that inertia could be interpreted also as a reaction force of interactions between the ZPF and quarks and electrons, the fundamental components of matter. This idea was improved with a further paper of (Rueda and Haisch), avoiding the previous *ad hoc* particle-field interaction model (Planck oscillator) as well as the initial formula complexity.

The author adopted the very likely idea that inertia and gravity are interrelated and found a scenario that is furthermore able to interrelate also EM-fields, the "Tunnel Effect" and the still uncertain “antigravitation” - if it does exist. For our proposes, we can consider the BF as a 3-D field of virtual gravitons inside an absolute void space (what we call a "perfect vacuum").

This model leads further to the prediction of holes in the BF, where the velocity of any particle can increase beyond "c", since such holes are lacking the inherent resistance (inertia) of the surrounding BF. Holes in the BF have the same properties as the "Tunnel-Effect" (overlight speed) and are generated by the same means (intense EM radiation). The "Tunnel-Effect" is therefore the first evidence for the existence of the BF, although holes in the BF allow a *real* overlight speed, and not only a *group speed.*

A second evidence is the higher temperature of the solar corona (up to 2x106°C) with respect to the "colder" photosphere (5,500°C). This could be explained by the high pressure of radiation that just leaves the material surface of the sun and produces large "holes" in the BF. In such holes, the inherent resistance (inertia) decreases, and further photons and particles that are continuously released by the photosphere can subsequently accelerate beyond the speed of light, thus gaining more energy and making increase the temperature of the solar corona.

To understand this phenomenon, we must consider that - because of the great material density of the sun - a photon is known to need approx. 1 million years to leave the body of the sun. This is because photons are again and again thrown back towards the center of the sun due to the enormous number of interactions that take place inside the sun. But once at the border of the photosphere (the last material layer of the sun), a dense beam of photons and particles is irradiated, and the high density of these photons produces holes in the BF by literally pushing away the field lines of this ground field. The result is that additional solar photons and particles do no longer find any resistance to their movement, and are therefore able to accelerate beyond "c". This *real* acceleration produces consequently an energy increase that makes in the end increase the temperature of the corona. Finally, as the photons and particles move away from the corona, their density decreases again with the square of the distance, and the holes in the BF consequently disappear. The final result is that outside the corona, photons adopt again their usual velocity ("c") and the ambient temperature consequently decreases to "normal" values.

Results

We can imagine our universe as consisting of a primary absolute void (a space with no resistance) and a BF that consists of virtual gravitons (VG) linked together by strings, thus building a 3-D matrix of such VGs that generates the effects we are all familiar with, like inertia, gravity, EM-fields, "Tunnel Effect", etc. Without the BF, our universe would be completely different, since particles would accelerate beyond "c" as there would be no longer any inherent resistance (BF) that produced inertia and gravity.

The BF must consist of VGs in order to be able to produce gravitation. VGs must further consist of strings in order to be able to produce inertia by the tension of the strings. The result is that the BF might consist of a 3-D matrix of VGs, where each VG is linked by strings to other 6 VGs (up, down, front, back, left, right = + and - values of x, y, z axes). The rows and lines of such VGs and strings represent the field lines of the BF.

The BF is able to generate EM and gravitational fields by means of the following interactions with material particles:

1. NEUTRAL INTERACTION

The BF would be eternal in absence of particles, but in our universe, it changes constantly due to the overall presence of material particles (fermions). When a neutral fermion moves, it interacts constantly with VGs of the BF. One part of the kinetic energy of the fermion is hereby transferred to any interacting VG on its trajectory. For any interacting VG of the BF, one real graviton (RG) is built (gravitation wave). In consequence, any moving fermion is loosing constantly kinetic energy and producing gravitation waves.

A punctual fermion interacts always with only one VG at a time. If such a fermion has a kinetic energy E_k, any RG that is produced by interactions would have the potential energy of a VG of the BF, plus a minimal kinetic energy that is necessary to loosen the 6 strings that anchor the VG in the BF:

[1] $E(RG) = E(VG) + E_{kmin}$

Where:
- $E(RG)$: Potential energy of a produced RG.
- $E(VG)$: Potential energy of an interacting VG.
- E_{kmin} : Minimal kinetic energy of a fermion, necessary to loosen the 6 strings that anchor a VG in the BF.

The above minimal kinetic energy is therefore equivalent to the potential energy of 6 strings:

[2] $E_{kmin} = 6\,E(S)$

Where: $E(S)$: Potential energy of a string

In order to overcome the force of the 6 strings that anchor a VG in the BF, according to [1], it is necessary to apply a minimal force. This force corresponds to the **inertia** of a punctual particle since it represents the smallest possible **resistance of the space**:

[3] $F_i = 6\,E(S)/l = E_{kmin}/l$

Where: F_i : Inertia of a punctual fermion
l : Minimal length, a minimum force must be applied, in order to loosen the 6 strings of a VG

The minimal length „l“ corresponds probably to a value close to Planck's Elementary Length, since the length of a string is probably the smallest length that can exist at all.

The constant interactions with VGs withdraw kinetic energy from a moving punctual fermion, so that its kinetic energy becomes always less. With each interaction, according to [2], a particle looses the potential energy of 6 strings:

$$[4] \quad E'_k = E_k - 6\,E(S) = E_k - E_{kmin}$$

Where: E'_k : Kinetic energy of a punctual fermion after a neutral interaction.
E_k : Kinetic energy of a punctual fermion before a neutral interaction.

A neutral fermion is therefore constantly decelerated by the inherent resistance of the BF. Furthermore, there is a minimum velocity at which a neutral fermion can move through the BF. To achieve this minimum velocity from the “absolute rest“, it is necessary to apply a minimum force in order to overcome the resistance of the BF. This force is again the **inertial force**. A particle can move through the space, only if it is accelerated to the above mentioned minimum velocity.

Supposing a punctual fermion is in a certain instant overcoming inertia from the absolute rest, it will achieve a minimum velocity, interacting each time, in a minimum time, with 1 VG, at a length, approx. equal to Planck's Elementary Length:

$$[5] \quad F_i = m\, v_{min}/t_{min} = 6\,E(S)/l$$

Where: F_i : Inertia of a fermion.
m : Mass of a fermion.
v_{min} : Absolute minimum velocity that a fermion can reach in the BF.
t_{min} : Minimum time, necessary to loosen the 6 strings of one VG in the BF.
l : ~ Planck's Elementary Length.

Each time a VG from the BF interacts and produces a RG, the BF changes (it **contracts** due to the expulsion of 1 RG out of the 3-D matrix). When a fermion crosses the space, the global contraction of the BF is proportional to the total amount of interactions with VGs of the BF. As a result, in our universe, the BF is constantly contracting and a constant **momentary reduction of the BF** takes place. Since in the space, there is an almost unlimited number of VGs, the BF is reorganized constantly by the surrounding VGs. To do this, the free ends of the strings of those VGs, adjacent to the VGs that were converted into RGs and left the BF, do connect each other, thus producing again an intact, although contracted, BF.

2. ELECTROMAGNETIC INTERACTION

According to convention, the field lines of a positive charge are directed outwards (out of the charge), while the field lines of a negative charge are directed inwards (into the charge). This can be explained as follows: a positive charge is constantly interacting with VGs of the BF, exciting and converting them into virtual photons (VPs) that are radiated in every direction. This radiation produces field lines that are directed outwards the positive charge. (In field lines of EN-fields, VPs are linked together by strings as in the BF.)

On the contrary, a negative charge interacts with VPs from surrounding EM fields and converts them back into VGs that can get linked again to the BF, thus reducing the constant reduction to the BF to a minimum. The direction of the field lines of an electric field is therefore equivalent to the flow direction of the VPs in the field, so that the field lines of a negative charge are always directed towards the charge.

A positive charge interacts with VGs of the BF, thus producing VPs that build up an electric field. The stronger the positive charge, the more energy the produced VPs have and the stronger the resulting electric field is:

[6] $E(VP) \sim q(+)\, E(VG)$

Where: E(VP) : Potential energy of a produced VP.
q(+) : Positive charge of an interacting particle.
E(VG) : Potential energy of an interacting VG.

For this reason, the charge of a particle and the strength of its electric field are two different phenomena. Without the BF, positive charges would not be able to build electric fields. They would be potentially positive, but effectively neutral, since they could not interact with VGs and produce VPs of the corresponding electric fields.

On the other hand, a negative charge interacts with VPs from surrounding EM fields, thus absorbing their potential energy and converting them back into VGs that can link again to the BF or interact with surrounding positive charges. Analogous to [6], the potential energy of a produced VG is here indirectly proportional to the negative charge of an interacting particle because, as known, VGs have less energy than VPs:

[7] $E(VG) \sim E(VP) / q(-)$

Where: E(VG) : Potential energy of a produced VG.
E(VP) : Potential energy of an interacting VP.
q(-) : Negative charge of an interacting particle.

In summary, the total field strength of an electric field is directly proportional to the number of VPs that build the field and to their individual potential energy:

[8] $F_e \sim n\, E(VP)$

Where: F_e : Electric field strength.
n : Number of VPs that build an electric field.
E(VP) : (Mean) potential energy of a VP in an electric field.

In conclusion, a charge can build an electric field, only if the space around the charge is full of virtual particles of the BF or of EM-fields. VPs emitted by positive charges interact according to this model with negative charges that convert them back to VGs of the BF. Therefore, our universe is a great electromagnetic circuit in balance.

Since the EM force is known to be approx. 10^{41} times stronger than gravitation, a VP of an EM-field must therefore have approx. 10^{41} times more potential energy than a VG in a gravitational field:

$$[9] \quad E(VP) = 10^{41} E(VG)$$

Where: E(VP) : Potential energy of a VP.
E(VG) : Potential energy of a VG.

This means that the potential energy (tension) of a string in an EM field is approx. 10^{41} times higher than that of a string in a gravitational field. As mentioned above, decisive for the spinning direction of an electric field is the direction in which the VPs of the field are moving (and not that of the VGs). Finally, since positive charges interact with VGs of the BF and convert them into VPs, this model includes the fundaments of what has been called **electrogravitation** by linking gravitation and electromagnetism (see chapter 3. below).

A. Electric attraction and repulsion

Two equal charges repel mutually because of the tension of the strings that link the VPs in the two fields:

Two positive charges repel mutually because each positive charge produces VPs as seen above. Since VPs are linked together in EM fields by means of strings, and strings in EM fields are relatively strong according to [9], VPs cannot change from one field line to another. Therefore, if we try to approach two positive charges, the strings that maintain the VPs linked together in each field produce a tension that tends to separate again both charges. This mutual repulsion is furthermore maintained by the VPs, both positive charges are constantly emitting.

According to [6], the stronger the charges, the more potential energy the produced VPs and the corresponding strings have. The result is, that the repulsion between two positive charges increases with their charge.

On the contrary, two negative charges repel each other because each charge interacts with VPs from surrounding EM fields and converts them back into VGs. There is a constant flow of VPs towards each negative charge that build two neatly defined fields since VPs cannot change from one field to another due to their high energy. In consequence, there is a constant **competition** for VPs between two negative charges, that produces a tension between both fields and repels the charges mutually.

Any electric repulsion or attraction is therefore proportional to the total tension (e.g. potential energy) of all the strings in the corresponding electric fields. Since every VP is linked to the corresponding electric field by means of 6 strings:

$$[10] \quad F_q \sim 6\, n\, E(S)$$

Where: F_q : Attraction or repulsion force between two electric charges.
n : Total number of VPs in both electric fields.
E(S) : Mean potential energy of a string in both electric fields.

Two unequal charges attract each other because the positive charge produces VPs that can interact directly with the negative charge. In this sense, two unequal charges **support** each other and a flow of VPs from the positive to the negative charge takes place. There is also a flow of VGs from the negative to the positive charge that is probably only partial, since a part of the VGs might flow back to the BF due to their extreme "volatility".

Remembering that there is a flow of VPs from the positive to the negative charge, if we try to move away each other two unequal charges, the strings in the combined electric field get tensed and a **resistance** appears since we are working against the field that is attractive.

On the contrary, if we approach two unequal charges, we work in the same direction of the field, no tension appears and the charges attract each other. Since the tension of the strings increases if we move away two unequal charges each other, and it decreases in the opposite direction, two unequal charges always tend to attract mutually.

According to the above model, in a neutral atom, there is a closed circuit of VPs between protons and electrons. Protons absorb VGs from the BF and produce VPs that interact with electrons and are again converted into VGs. A positive ion has an excess of protons that produces an excess of VPs that leaves the atom along field lines. On the contrary, in a negative ion, the excess of electrons interact with surrounding VPs and a negative electric field appears by means of field lines of VPs that enter into the atom. In consequence, an atom is electrically neutral if there is no interchange of virtual particles with the surrounding space. Furthermore, without the BF, any positive charge or ion would be electrically neutral since there would be no VGs to interact with and no VPs could therefore be emitted. Negative charges would also be electrically neutral in this case, since there would be no VPs to interact with.

3. GRAVITATION

The BF is no gravitational field by itself. Gravitation happens only if there are fermions with certain kinetic energy inside the space. Since in our universe, every particle is in a constant movement, together with all the existing galaxies and other celestial formations, fermions are therefore constantly interacting with VGs of the BF. A part of the kinetic energy of fermions is transferred to VGs, and according to [1], RGs are produced in form of gravitation waves.

For each interacting VG that leaves the BF as a RG, there is a **momentary reduction of the BF**, and the BF consequently contracts. A second particle close to an interacting particle is pushed consequently closer to the latter. Therefore, it seems to exist a force that tends to attract both particles mutually, but in effect, it is the BF that has become contracted between both particles, thus producing the “illusion” of a gravitational attraction. In consequence, a constant momentary reduction of the BF due to the presence of moving fermions is what we call “gravitational field”.

A momentary reduction of the BF is able to approach two fermions mutually, since both are constantly interacting with VGs of the BF and are therefore constantly anchored to it. This is because the above mentioned neutral interactions loosen the strings that maintain VGs linked in the BF. In order to loosen a string, it is necessary that a fermion connects physically to a VG, and in this state, it is anchored to the BF and can be moved by momentary reductions of the same. Furthermore, any fermion is very large with regard to VGs, so that it always interacts with several VGs at the same time and never stops to interact.

The faster a fermion is, the more interactions with VGs will happen, and the more the BF contracts. In consequence, faster particles have a higher “gravitational attraction” than slower particles. In consequence, the gravitational attraction of a particle could be very small or zero if it did not move with respect to the BF. Therefore, this model predicts that matter in absolute rest (if this could be achieved) will present no gravity. Consequently, a further evidence for the BF would be if we could detect that matter cooled near the absolute zero has less weight than at higher temperatures.

4. MAGNETISM

According to this model, the flow of VPs in a magnetic field is different to that of an electric field since there are no magnetic monopoles. The smallest possible magnet is the atom or molecule. As we know, the field lines of a magnet run inside the magnet, from the north pole to the south pole, and outside of the magnet, from the south pole to the north pole (according to this model, there is a flow of VP in exactly these directions). An internal cyclic current (see for ex.: De Curtis, pg.56) transfers energy to the VGs of the BF, which are in this way converted into VPs of the resulting magnetic field.

Since, as in electric fields, VPs are linked in EM fields by strings, they cannot move from one field line to another. In this sense, if we approach two equal magnetic poles, the field lines cannot interchange VPs, so that a pressure on each field line appears and the strings between the VPs of the fields become tensed. This tension is proportional to the repulsion force between two equal poles. The more we approach two equal poles, the smaller becomes the distance between two adjacent field lines and the higher the tension of the corresponding strings is. The result: equal poles tend to repulse each other and this repulsion is proportional to the distance between both poles, e.g. between two adjacent field lines.

If we approach two unequal poles, the VPs that come out of the south pole of one magnet enter without any impediment into the north pole of the other magnet. To do so, it is not necessary that VPs change from one field line to another. The density of field lines between two opposite poles simply doubles and a unique circuit of VPs appears. Since the combined magnetic field consists of twice as much field lines (and VPs) as in one single pole, the magnetic force of the combined magnet is also approx. twice as high as that of one individual magnet.

In magnets, without the BF, internal currents would not be able to produce VPs, since there would be no VGs to interact with. In this case, there would be no magnets at all, as well as no positive charges or ions. In addition, there would also be no negative charges or ions since the VPs with which they interact would not exist. In conclusion, without the BF, there would be no EM fields at all.

Discussion

The BF is probably the principle that makes our universe be as it is. We are not able to "see" the BF, because we are submerged in it. This case is similar to a diver who is not able to see the surrounding water or a man who is not be able to see the surrounding air. But on land, we are able to see water drops on our hand and if we abandon the earth, we are also able to see the atmosphere. It is not possible to see a medium if we do not leave it, because of the **lack of contrast**. But once we have abandoned our medium, we can see it because of the greater contrast with other existences. The same happens with the BF: since it is the medium in which we are all submerged, we cannot "see" it because any physical activity in our universe is due to the presence of this medium and would not take place without it. The surrounding nature would not be the same without the BF and we are not able to imagine a world without it. In order to "see" the BF field, we might increase the contrast, e.g. we might "step out" of the BF (see chapter 6. below).

Analogous happens with virtual bosons. Virtual bosons are linked together by means of strings, thus building part of a medium (field), we cannot see. But if a virtual boson abandons the medium (i.e. as free photon), we are immediately able to detect it by means of technology (radio waves) or our eyes (light) because of the contrast. In consequence, in our universe, there are at least 4 different fluid media (from more to less density): liquids, gases, the "perfect vacuum" (BF) and the "absolute void" (nothing) (see chapter 6. below).

Experimental evidence for the presence of the BF can be obtained, since the speed of light depends on the density of the BF. Since the BF is more dense near celestial bodies (gravitation), the speed of light should increase in direction to the outer space. We should be able to detect an increase with respect to a fix length (like a ruler). This increase cannot be detected directly (by means of simply emitting a photon beam and counting the time it takes to reflect) because the lower density of the outer space makes "enlarge" the space parallel to the decrease of density of the BF. But a fix object (ruler) is not affected by this phenomenon and can be used as reference. The above term "enlargement" means that the field lines of the BF become less dense, so that a photon can reach higher speeds by simply passing through the greater empty spaces between the field lines since these spaces (absolute void) do not have any resistance.

In summary, the BF would be in the end responsible for (and would definitely explain) a great lot of physical phenomena, many of them controversial, including:

1. INERTIA

Inertia is the resistance of the BF that limits the freedom of movement of any material particle. This resistance is due to the interaction between VGs of the BF and fermions. Each fermion has a certain kinetic energy, and a part of this energy is transferred by means of interactions to VGs of the BF. As a result, it is necessary to apply a force in order to move a fermion through the BF from the absolute rest in order to compensate the kinetic energy, the BF absorbs. This force proportional to inertia and is due to the potential energy of the strings that link VGs together in the BF ([3], [5]). For each interacting VG, a fermion must loosen 6 strings in order to emit a RG.

2. THE FALL OF THE BODIES

Any body falls to earth always with the same velocity, independently from its mass, supposing there is no resistance of the air. Since the field lines of the gravitational field are directed vertically towards the center of the earth (see also chapter 3. "GRAVITATION" in Results), every fermion in a body moves during a free fall along these field lines. Each field line consists of numerous VGs that are linked together by strings. For this reason, any fermion must interact with a certain amount of VGs on each field line in order to fall to earth. It does not matter how many fermions a body has, since any fermion must realize these interactions **independently** from the other particles of the body, because any particle is located on an individual field line (or group of field lines). As a result, each fermion falls at the same time to earth, independently in which body it is momentarily located. This means furthermore that each body falls at the same time to earth, independently from how many fermions it is made of.

3. ANTIGRAVITATION

In addition to an electric field, charges do also have a gravitational field due to their mass. Positive charges interact with VGs of the BF. Therefore, VGs are necessary to build up the gravitational <u>and</u> the electric field of positive charges. In consequence, there is a **competition** between both fields for VGs, so that the stronger field (electric field) weakens the weaker field (gravitational field). The gravitational field does no longer dispose of 100 % of the VGs of the BF to be built up, with the result that it becomes weaker. This is the above mentioned **electrogravity** effect.

Since any VP in an EM field signifies a VG less in the corresponding gravitational field, the decrease of gravity of a body is proportional to the amount of positive charges of the body that determine how many VGs are converted into VPs of the corresponding electric field:

[11] $(-)F_G \sim n\, q(+)$

Where: $(-)F_G$: Gravity decrease of a body.
n : Number of positive charges of the body.
q(+): Load of a positive charge of the body.

This type of antigravitation (EM reduction of gravity) happens only with <u>positive charges</u> and can reach theoretically a value from 0 - 100 % according to how many VGs of the BF the positive charges interact with (absorb).

<u>Negative charges</u>, on the contrary, emit VGs, so that in this case, there would be no deficit of VGs that could produce antigravitation. In any way, VGs produced by negative particles are very "volatile" and cannot always interact with nearby positive particles (for ex. inside an atom), so that they escape partially to the BF without interacting. This produces always a slight

deficit of VGs close to negative particles and thus, always a slight local reduction of gravity. In this sense, also negative particles would participate in antigravity, although not directly.

Antigravitation was found accidentally in an experiment with a turning superconductor that was suspended by solenoids (Podkletnov). In agreement with the experiment, antigravitation cannot be suddenly be switched on and off. It reaches a theoretical value up to a certain percentage. In the experiment, it was up to 2.1 %. This can be interpreted as if approx. 2.1 % of the VGs of the BF had been converted into VPs of the corresponding EM field, so that gravity had decreased in exactly this proportion.

The supposition that VPs in EM fields must not necessarily be very abundant with respect to VGs in a gravitational field, is supported by the considerations in chapter 4. „MAGNETISM“: if we approach 2 unequal poles, the field lines of both magnets combine to one unique field. This is an indication that these field lines are not so abundant as in the BF. Otherwise, such type of addition would probably not be possible because of the lacking space between two adjacent field lines (in the BF, this space is probably close to Planck's Elementary Length). Therefore, the above supposition that it is about 2.1 % of the total virtual bosons, is effectively a realistic idea.

4. THE SPEED OF LIGHT

It is still unknown, why the speed of light has a finite value. This model gives a possible explanation: photons move through the interstices that exist among strings and VGs in the BF. On one hand, collisions with VGs limit the speed of photons. On the other hand, the lack of resistance of the interstices tends to increase the speed. The logical result is a movement with a limited velocity. Without the BF, the speed of light would be probably infinite (see also below).

Furthermore, light could be wavy because photons do oscillate around the VGs of the BF. As a result, without the BF, light would probably be linear and of practically infinite speed.

5. HOLES IN THE BACKGROUND FIELD

According to [3], the resistance of the space is equal to inertia. Since this resistance is based exclusively on the presence of VGs in the BF, it is easy to imagine that, if there was no BF, the resistance of the space would be equal to zero and there would be no inertia. Consequently, particles would cross such a void space with an infinite velocity, from one end to the other, as if there was a "hole in the universe".

A hole in the BF would also mean a complete interruption of fields of force, since virtual bosons in fields of force are linked together by strings. The considerable tension between these strings would probably avoid that a part of the field of force "evaporates" in the absolute void of the hole. On the other hand, fermions and free bosons (i.e. light particles) could pass through such holes easily because they are not linked to any field by means of strings.

In consequence, if astronomers detect certain galaxies or celestial formations that do not influence each other mutually according to the known laws of physics, this could mean that there is a hole in the BF between these formations. In such a case, the mutual attraction, as well as any electric and magnetic field, would be interrupted by the hole.

6. THE "TUNNEL EFFECT"

The Tunnel Effect is per definition the propagation of EM waves with overluminic velocity and has been described experimentally, for ex. in (Nimtz). In this sense, this phenomenon is probably a small variant of the above described holes in the BF since both phenomena have the same effect (overluminic velocity). In consequence, it is logical to assume that both phenomena do also have the same cause (holes in the BF).

Nimtz used intense EM waves to increase "c" in hollow conductors. This could be explained by the fact that EM waves literally push away field lines of the BF, thus producing the mentioned holes in the BF. Further photons could then pass such holes with a velocity higher than "c".

Of course, such artificial tiny holes in the BF are probably not perfect, so that there will always be a certain resistance. Therefore, particles will probably need a certain time to accelerate from their ground speed to overluminic speed. For this reason, in small holes, we cannot expect very high velocities. And in effect, a typical value that has been measured here is only 1,7 c (Steinberg et al.) or 2c (Nimtz), thus supporting the idea of imperfect holes in the BF.

Conclusions

We have seen that the idea of the BF meets very well the results of already made experiments and the principles of physics. It cannot be foreseen, which possibilities we are offered if we learn to produce and to control holes in the BF. A practical application could be for ex. instant teleportation of material bodies. Through holes in the BF, we could also receive signals from remote civilizations and send them our own signals. We would dispose in this case of the possibility to communicate, even without any physical contact. In this sense, information has already been tunneled without being destroyed (Nimtz).

In (Calvet), further practical applications of the BF-theory are discussed, such as the possibility of interstellar voyages with the aid of field and gyroscopic propulsions. Overlight speed could be easily achieved by means of a concentrated beam of EM-waves that produces a hole in the BF. Spacecrafts could use such technology to travel even to distant galaxies.

Another possibility is the transmutation of elementary particles, since the model of the BF allows theoretically to extract single photons from EM fields in an analogous way as gravitons emerge from gravitational fields.

Since any momentary reduction of the BF signifies that the BF is absorbing energy and that it is vibrating, it should be possible for us to use these vibrations (zero-point energy) so that we could dispose of a practically unlimited amount of cheap energy due to the practically unlimited number of interactions between fermions and the BF in our universe.

The idea of overluminic speed is furthermore no longer contradictory with Special Relativity, since Special Relativity is based on a universe that contains a BF. Therefore, if we make a hole in the BF, the speed of a particle can become practically infinite, even in agreement with Einstein's principles.

Acknowledgements

I wished to tank all those people (like my wife Maria) that have supported during all these years my experiments of thought, which lead gradually to the BF-theory. I hope this theory will be very useful for mankind and that we learn with its aid more about the real nature of our universe.

Literature

C. Calvet, „Raumfahrzeuge der Zukunft", Bohmeier Verlag, Leipzig, 105 pp. (July 2000) (in German)

S. De Curtis, J. Fernández Ferrer, „Physik", Neuer Kaiser Verlag, 95 pp. (1998) (in German)

B. Haisch, A. Rueda and H.E. Puthoff, "Inertia as a Zero-point field Lorenz force", Phys. Rev. A **49**, 678 (1994)

G. Nimtz, „Instantanes Tunneln - Tunnelexperimente mit elektromagnetischen Wellen"" Phys. Bl. **49**, p. 1119-1120 (1993) (in German)

E.E. Podkletnov (Moscow Chem. Scientific. Ctr.), "Weak Gravitation Shielding Properties of Composite Bulk $YBa_2Cu_3O_7$-x Superconductor Below 70°K under E.M. Field," Univ. Cincinnati Engineering, report # MSU-chem 95, abstract cond-mat/9701074, 1997; 19 pp., in: NEN (New Energy News), **Vol. 4, No. 12**, p. 7-26 (April 1997)

A.Rueda & B.Haisch, "Inertia as reaction of the vacuum to accelerated motion", Physics Letters A, **Vol. 240, No. 3**, p.115 (1998)

A.M. Steinberg, P.G. Kwiat & R.Y. Chiao, „Measurement of the Single-Photon Tunneling Time", Physical Review Letter **71**, p. 708-711 (1993)

PART 2

Journal of Theoretics

DETECTION AND ORIGIN OF THE "BACKGROUND FIELD"

Author: Carlos Calvet, Ph.D.

Dr. Carlos Calvet, Francisco Corbera, 15; E-08360 Canet de Mar (Barcelona), Spain

E-mail: planckworld2@terra.es Web: personal.telefonica.terra.es/web/planckworld/

Abstract: The Background Field Theory (BF Theory) was able to explain the nature of inertia, as well as of that of gravitational and electromagnetic fields. Later, it was revealed to even predict the existence of so called "Torsion Fields." This would be the first time that all these phenomena were unified in a unique model. Furthermore, it has been recently revealed that Russian astronomers may have already detected the BF during the late 70's as a radiation or source that is able to produce gravitation. More than 20 years later, this author developed the model of the BF while the Russian evidence of such was still under secrecy after the Cold War.

The BF theory makes superfluous the need for time or chronological effects to explain phenomena such as "non-locality" or "entangled particles" because of the unlimited velocity that particles may have in the "diluted" regions of the BF. In the synthesis of the Big Bang theory and BF theory, the origin of the BF has been assigned to a so-called "Primordial Star" (PS), produced by the implosion that might have accompanied the explosion of some initial particle during the Big Bang. Such a PS would be still inside our universe and could be used as "cosmic lighthouse" for future interstellar navigation.

Keywords: Background Field, Big Bang, EM fields, gravitation, inertia, strings, Torsion Fields

Introduction

Probably the most unknown of all physical aspects is inertia, the instantaneous opposition to the acceleration of material objects. It was formerly believed that inertia was the result of the motion from all of the matter that exists in the universe (Mach's Principle), but this belief has never been formulated successfully[1] and has even been found to be false[2]. Furthermore, the experiments of Hughes and Drever[2] showed that "the entity that produces inertia, if other than matter itself, displays a great degree of isotropy".

Recently, Haisch, Rueda & Puthoff proposed that "inertia could be interpreted, at least in part, as a reaction force originating in the interactions between the electromagnetic Zero-Point Field (ZPF) and the elementary charged components (quarks and electrons) of matter."[3] Rueda & Haisch developed a more simple concept that "would substitute Mach's principle and imply that no further mass-giving Higgs-type fields may be required to explain the inertia of material objects."[4]

An additional approach to the still unknown origin of inertia was made by Calvet[5], the author, who proposed the existence of a so called "Background Field" (BF), the origin not only of inertia, but also of gravitation and electromagnetic fields, as well as being responsible for cosmology and all its implications.

The concept of the BF was created and based on the most ample pure logic. It should include, an explanation in plain words and not contradict General and Relativistic Physics as well as Cosmology and especially, the above-mentioned findings (isotropy, ZPF, etc.). The result was the creation of a model that effectively predicted the existence of an absolute isotropic field, the Background Field (BF), located inside an *absolute* void space. In this sense, the BF would consist of virtual gravitons (VG) linked by strings (derived from common String Theories), so that contractions of the BF would generate gravitational fields, rotation would generate electromagnetic fields, while in the so called "holes" of the BF, the absolute void beneath the BF would emerge, thereby allowing particles to practically accelerate to an infinite speed.

This latter property of the BF made obsolete the concept of "time," when trying to explain how present, past, and future are interconnected[6,7], how a particle could be in different places at the same time[8] and even with regard to the properties of so called "non-locality" or "entangled particles"[9,10,11] since an unlimited velocity would allow any particle to be in instantaneous contact with the rest of the universe without any delay. The above properties of "non-locality" and "entangled particles" seemed to demonstrate that, under certain circumstances, "there is nothing to prevent *instantaneous* interactions between particles, no matter the distance between them," and is furthermore the principle of action of the BF as explained in Calvet.[5]

In addition, the BF theory explains in a simple way the existence of so called "Torsion Fields" (TF), as introduced by Shipov, Spasskii & Moskovskii, and others. To understand how the BF originates TFs, we must remember that any conventional elementary particle has a spin (electrons/quarks = 1/2, photons = 1, gravitons = 2) and that since material particles are constantly spinning around their own axis, they constantly interact with the BF, since they are submerged in it. While this interaction produces (according to Calvet[5] and Calvet[12]) gravitation waves, interactions with bosons do not give place to the emission of any particle or wave. Instead, their ground linear movement turns into an oscillation around the virtual particles of the BF, so that we observe boson waves instead of boson beams in our universe (the wavy nature of bosons). Fermions, on the contrary, interact physically with virtual gravitons of the BF, so that the spin of fermions is transferred to the BF, making a region close to the interaction zone spin harmonically with the fermion, thus building an elementary Torsion Field (TF).

In this sense, TFs might be of any size since they could be produced by any kind of spin, rotation, or orbit. So, there are possible various types of TFs, from smaller to larger such as that produced by elementary particles (fermion spin), atoms (electron orbits), molecules and macroscopic objects (i.e., a turning wheel or molecule), moons and planets (orbits around planets or a star), or even galaxies (rotation of a galaxy).

After stating that the BF theory explained very well all that was mentioned above and also strange additional phenomena, like antigravity[13] and superluminous speed, without destructive effects[14]; more evidences and effects of the BF were shown in Calvet.[12] It was suggested among other things, that the "extremely high temperature of the solar photosphere (up to $2x10^{6}$°C) with respect to the material surface of the sun (approx. 4000°C) could be explained by the high pressure of radiation that just leaves the surface of the sun and produces large "holes" in the surrounding BF. The lack of resistance in such holes would allow photons to accelerate beyond "c," thus gaining more energy and causing an increase of the temperature in the corona".

In his first popular scientific book, Calvet[15] developed, in part independent of the BF theory, propulsion systems for spacecrafts that could be able to navigate through the outer space, while fulfilling 2 of the 3 main principles of NASA for future spacecraft which are: 1) travel with practically no fuel and 2) travel with a speed that could approach or even exceed that of the speed of light. The 3rd principle (new energy sources) was found to be irrelevant, since the available sources (and even Newtonian physics) already allow principles 1 and 2. In addition, when applying the principles of the BF theory, a spacecraft could even navigate in a similar manner as an elementary particle in outer space.

Detection of the Background Field

To the surprise of this author, it was recently revealed that already in the late 70's, Russian scientists had found a field or energy source that had effects exactly equivalent to those predicted by the BF. Such a field that exists independently from matter and other energy sources and that in its original form (a 3D net of virtual gravitons) would be more evident in the outer space than near galaxies, since in presence of matter it interacts becoming a gravitational field (thus emitting gravitation waves).

In October 1977, the Russian astronomers Kosyrev and Nasonov performed measurements of several stellar objects, using for the first time a crystal detector grown outside the gravitation field of the earth. They wanted to measure the detailed intensity profiles of some objects, in order to verify astrophysical models. It was expected to find maximal radiation intensity in the center of the galaxies, while the intensity towards the periphery was expected to decrease exponentially in agreement with the lower star density in these regions.

But contrary to the expected scenario, the new detector provided an intensity profile, that was exactly *contrary* to the light profile expected, e.g. with a *minimum in the center of the galaxies and a maximum towards the boarders*. Even with the reflector covered by a metal plate, the signal they received remained unchanged. After shielding and compensating the detector in different ways, the effect could not be eliminated and was even stronger in certain cases.

After years of further research in an attempt to seek a logic explanation, they concluded that the measured effect was definitely gravitation. Ample laboratory experiments they performed have confirmed that the detector was extremely sensible to gravitation. But contrary to any known gravitation theory (e.g. Newton, Einstein), the effect was not stronger close to matter (galaxies), rather it was stronger further away in the outer space.

They suggested that "if gravitation is anti-symmetric to electromagnetic waves (light), the only explanation that remains is that gravitation is a medium, that is generated in space and absorbed by matter. Matter is therefore not an emitter of gravitation, but a 'collector.' Probably, at the marginal regions of the observed galaxies, gravitation is being converted into matter, and closer to the stars, gravitation is probably absorbed strongly, so that here, a deficit of gravity appears".[16]

From 1978 to 1979, the detector worked in a semiautomatic way and detected twin and even triple intensity profiles. After eliminating all possible errors, the authors stated that each galaxy had up to three profiles that were practically identical and shifted mutually to a constant distance. Astronomic calculations revealed that this distance corresponded exactly to the way the galaxy had run during the time it took the light from the galaxy to reach the earth. They then directly deduced firstly that the profile (left to right) corresponded to the position of the galaxy at the moment t=t0-r/c (e.g., in the past); secondly, to its current position t=t0 (the present); and thirdly, to the future position of the galaxy at the moment t=t0+r/c, where "r" is the distance from the galaxy to Earth.[16]

Finally they concluded that "even if the future cannot be calculated, it could be observated".[16] But even if some of their findings met the geometry Minkowski-space (a space, where past, present and future can coexist), their observations contradicted Special Relativity. Therefore, they were forbidden to continue their research, even though they had been able to determine for the first time in history, the complete velocity vectors for galaxies in a 3D space.

Conclusions Drawn from the Kosyrev & Nasonov's Findings

Firstly, we observe that the above findings are completely coincident with the predictions of the "Background Field (BF) Theory":

1. There exists an isotropic field in the universe (the most isotropic of all fields is a field with the properties of the BF).

2. The isotropic field is independent to the presence of matter [Calvet[15]].

3. The isotropic field is able to generate gravitation [Calvet[5] & Calvet[12]].

4. The isotropic field weakens in presence of matter (as it produces gravitation waves that "dilute" the BF [Calvet[12]]).

5. The effects of the isotropic field cannot be shielded by metal plates, as in the case of light (since it is "hyperfluid" [Calvet[15]]).

Secondly, we observe that there were 3 copies of the signal profile found, shifted mutually by a concrete value. Though this phenomenon has been explained with the aid of stationary waves, chronons, or even Torsion Fields, there are other easier explanations that are not completely impossible as discussed in the following paragraph.

Thirdly, we observe that the cited crystal detector was grown outside the reach of the gravitation field of the earth. This means that the structure of the crystal was not affected by gravitation so that it would be (even from a theoretical point of view) an excellent detector of gravitation and any kind of predecessor field that is able to produce gravitation (in this case, the BF). In this sense, if the crystal is such a good detector of gravitation, it is not negligible to suppose that it has effectively detected the BF (in outer space) and the dilution effect of boson waves on the BF closer to galaxies (since matter produces "BF-diluting" waves [gravitational, EM] while interacting with the BF).

But while the intensity increase of the signal towards the outer space indicates that there must be effectively some kind of isotropic field that produces such signals (the BF), the presence of a triple intensity profile could be theoretically, even a simple artifact, created by the measuring methods that were all very similar. In this sense, and if no other simple explanation can be given to the coexistence of 3 identical shifted profiles, I suggest to await more concrete data and different experiments that will probably allow us to explain this finding in a simple way, even if the given scenario seemed to be "out of the reach" of classical physics.

The fact that the 3 profiles found did not have the same intensity, seems to suggest that it is only one profile that has been reflected into 3 by some kind of reflection between the crystal and the galaxies. A good way to demonstrate that the are no time effects, would be to use a much more sensible detector. If this detector receives not only 3, but *more* copies of a main signal, this would definitely mean that the copies of the signal are being produced by conventional reflection and not by exotic "time or chronological effects".

Finally, a triple signal could also be produced in a natural way by objects with a halo or two or more rings as is usual in neutron stars and many other celestial formations. In this case, the central body of the star would produce the main signal, while portions of the halo or a ring to the left and a ring to the right would produce 2 additional faded signals that, under certain circumstances, would be very similar to the main signal. There are many other possibilities in explaining the triple signal and I suggest it is better not to take the most fantastic one.

Origin of the Background Field

Like any other field, the above mentioned BF ought to have a source. In this sense, in Calvet[15], a new cosmic concept is introduced: the "Primordial Star" (PS).

The PS is a concept that revealed to be evidently missing in any cosmic theory. If we take for example the most common theory, the so called "Big Bang Theory", known by all of us, we see that there seems to have been an enormous explosion of some kind of dense initial particle, and that this explosion, approx. 11 billions of years ago, seems to have generated all the matter in the universe, probably together with the corresponding space.

Even if there are severe doubts that the Big Bang has ever taken place,[17,18] mainly because the Red Shift and the Cosmic Background Radiation could be explained by the fading of photon energy due to the proven interaction of light and charged particles in the universe[19], the presence of the universe still ought to be explained in some way, even if we suggest that it began with the 'explosion of a soap bubble in the bath of some superior being'. And, even if the Red Shift and the Cosmic Background Radiation proved to be no good indicators for an initial Big Bang, this does not mean that something like a Big Bang did absolutely not take place.

In this restricted sense, and if we review the Big Bang Theory, we find that it does not meet Newton's 3rd Law of Dynamics, because any explosion (i.e. Big Bang) might be accompanied by a parallel *implosion*. Therefore, if we accept in some way, that the universe has been created by some kind of explosion that has dissipated matter (galaxies), we must also accept that there might have been a parallel *implosion* that compressed the core of the original particle or star that exploded.

This process is very well known, for example in stars that, according to its initial mass, produce a residual body called White Dwarf, Neutron Star or Black Hole, depending on the initial mass of the original star. In consequence, the ancestral explosion of the particle or star that might have given place to the universe might have also produced an extremely compressed body as in the above-mentioned cases of conventional stars. And this ancestral body is equivalent to the "Primordial Star" (PS) in the BF theory. (Even if the Big Bang theory talks about an initial "particle" that exploded, it seems to me more reasonable to talk also about a "star" because of the enormous mass of the here considered body).

We might still be able to detect the above mentioned PS in our universe. Since the PS would probably be still placed near the very center of the universe, it could be eventually used as "cosmic lighthouse" for future interstellar navigation and may be, even as unlimited energy source, once detected (Calvet[15]).

In accordance to Bennett[21], light can be thought of as "evaporated mass" and mass as "condensed light" according to the transformation equation, $E = mc^2$. If we take this principle and extend it to the above-mentioned scenario with a recently created PS, we will come to the conclusion that a body like the PS should effectively evaporate little by little when located inside a completely void universe. This process would be analogous to the "evaporation process" of Black Holes[20] and would in addition be a very likely source of the "Background Field" (BF).

Since any particle (real or virtual) can be explained by strings, one important conclusion of the BF theory is that any kind of elementary particle has in principle the same origin and different particles are simply a different manifestation of the same ancestral string. This conclusion gives place to the possibility of transmutation of one particle into another (e.g. of matter into energy and of energy into matter) not only in an Einstein-like sense, but also in the sense that it might be possible to extract a virtual particle from a field of force and convert it into a real particle like a photon or even an electron. If humanity manages to achieve this goal, we would be able to manipulate the universe even far beyond the combined principles of nanotechnology and String theories.

In agreement to neutron stars or Black Holes as the result of supernovae explosions, we can imagine the PS as an extremely condensed body or star, with probably a mass of approximately 10% of the mass of the whole matter in the universe. The PS might spin around its own axis at a speed that could be even not comparable to the speed of light. Due to its extremely high mass, the PS might produce a complete punctual distortion of the surrounding space, so that it even would no longer be located inside our 3D universe, but inside a "bag" in the so called "Hyperspace" (a space with more than 3 spatial dimensions).

Opposite to Black Holes that are believed to be "punctual," the PS is probably even much less than this. It could have a positive size in hyperspace, but a negative size in our 3D universe, so that we would not be able to measure its size if we did not manage to leave our universe and enter the hyperspace. Once in hyperspace (if we could survive there), we would probably see the PS as a relatively huge spinning object that seems to challenge any known physics or even the purest mathematics. Its specific weight would be so enormous that one simple grain of the PS, placed inside a galaxy, would act as a huge Black Hole, absorbing the whole galaxy within fractions of seconds. The PS probably consists in a huge "ball of strings", where strings are so intensely condensed that they practically build a continuum. It would therefore be the most solid body that had ever existed in the universe. But, while aging, it would "evaporate" in the above-mentioned manner, thus producing a BF that allows us humans to live in an otherwise dead universe. As a direct result of the interaction of the BF and the matter dispersed by the Big Bang, other secondary fields like gravitational, EM and Torsion

Fields may appear, giving our universe its definite shape and properties.

We should be able to detect the exact position of the PS, since it would be the most energetic body in the universe. Similarly to pulsars, quasars and other extremely energetic stellar bodies, the PS should be spinning at a very high speed, while producing at the same time very energetic radiation. Since the most energetic of all known radiation are gamma rays and cosmic rays of still unknown origin, the presence of the PS could explain where cosmic rays come from and why a Background Gamma Radiation and even spontaneous gamma ray explosions exist in the outer space.

Finally, I wanted to mention here again the main consequences of the BF theory for future interstellar navigation and technology:

- The BF can be "diluted" by a flood of EM-radiation, so that inside the diluted regions, speeds beyond the speed of light are possible for any particle or body, even for spacecrafts.
- The PS, when detected, could be used as a "cosmic lighthouse," thus allowing future interstellar navigation, even to remote galaxies without loosing control of the route.
- Any elementary particle, no matter if real (bosons, fermions) or virtual (fields of force), can be considered as an association of more or less condensed and/or twisted strings, so that it is theoretically possible to convert one particle into another, no matter its nature is, provided we know how to manipulate strings.

Cited References:

1. W. H. McCrea, Nature 230, 95 (1971).
2. S. Weinberg, "Gravitation and Cosmology: Principles and Applications of the General Theory of Relativity" (Wiley, New York, 1972), pp. 86-88.
3. B. Haisch, A. Rueda & H. E. Puthoff, "Inertia as a Zero-Point Field Lorenz Force", Phys. Rev. A 49, 678 (1994)
4. A. Rueda & B. Haisch, "Inertia as reaction of the vacuum to accelerated motion", Physics Letters A, Vol. 240, No. 3, 1998, p.115
5. Calvet C; "The Background Field Theory", Journal of Theoretics, Vol.1, No.5, Dec 1999/Jan 2000.
6. G.I. Shipov "Teoriya fizicheskogo vacuuma." ("Theory of physical vacuum"), Moscow, 1993, NT-Center, 362 p. (in Russian).
7. B. Spasskii, A. Moskovskii, "Non-Locality in Quantum Physics," Sov. Phys. Usp., vol. 27, no. 4, 1984, p 273.
8. M. Chown, "All the World's a Time-Machine," New Scientist, 7 March, 1998, p.

38.

9. D. Boumeester et al., "Experimental Quantum Teleportation," Nature, vol. 390, 11 Dec 1997, p. 575.

10. M. Buchanan, "Beyond Reality," New Scientist, 14 March 1998, p. 27

11. T. Sudbery, "The Fastest Way from A to B," Nature, vol. 390, 11 Dec 1997, p. 551

12. Calvet C; "Effects and Evidence of the "Background Field"", Extraterrestrial Physics Review - EPR, Japan, Vol.1, No.3, March 2000.

13. E.E. Podkletnov (Moscow Chem. Scientific. Ctr.), "Weak Gravitation Shielding Properties of Composite Bulk YBa2Cu3O7-x Superconductor Below 70°K under E.M. Field," Univ. Cincinnati Engineering, report # MSU-chem 95, abstract cond-mat/9701074, 1997; 19 pp., in: NEN (New Energy News), Vol. 4, No. 12, April 1997, p. 7-26

14. G. Nimtz, „Instantanes Tunneln - Tunnelexperimente mit elektromagnetischen Wellen"" Phys. Bl. 49, 1993, p. 1119-1120 (in German)

15. Calvet C: " Raumfahrzeuge der Zukunft (Spacecrafts of the Future) ", Bohmeier Verlag (Lübeck), July 2000, 105 pp. (in German).

16. N.A. Kosyrev & V.V. Nasonov, „Properties of Time, Discovered by Astronomic Observations", Russian Academy of Sciences of the USSR, Moscow/Leningrad, Series "Problems of Space Research", 9th edition (1980) (in Russian).

17. Anastasovski, P., et al., "A New Approach to the Cosmic Red-Shift and to the Cosmic Microwave Sources," Journal of New Energy, Vol. 1, No. 2, Summer 1996.

18. W. C. Mitchell, "The Cult of the BIG BANG, Was There a Bang?," William C. Mitchell, Cosmic Sense Books, Carson City, NV, c1995, 240 pp.

19. B. Lehnert & S. Roy, "Extended Electromagnetic Theory, Space-Charge in vacuo and the Rest Mass of the Photon", World Scientific Series in Contemporary Chemical Physics - Vol 16, Publ. by World Scientific, River Edge, NJ, c1998, 160 pp.

20. S. Hawking, "A Brief History of Time ", Bantham Books, NY, 1996, 271 pp.

21. Bennett, C., "Light is the Ghost of Mass," New Energy News, Vol. 5, No. 2, June, 1997, pp. 8-9.

Other References Not Specifically Cited:

- H. Müller, "Schwerkraft-Signale aus dem All empfangen", Raum-und-Zeit, no. 104,

March 2000
- A.I. Veinik "Termodinamika rjealnyh processov" ("Thermodynamics of real processes"), Minsk, Nauka i Tehnika, 1991, 576 p. (in Russian)

Acknowledgements

I wanted to thank Hal Fox of the "Institute of New Energy" for providing information about some of the Russian authors as well as for his general support.

I hope that the BF theory is useful to other scientists, not only to redact articles, but also to achieve the ultimate goal of all of our efforts, e.g. to help NASA to build in short a spacecraft that is able to navigate through the galaxy or even through the whole universe in real time and come back to render ourselves information about other civilizations or distant places.